Marine Animals

FROM THE NORTHEAST COAST

written & illustrated by
J. ROACH-EVANS

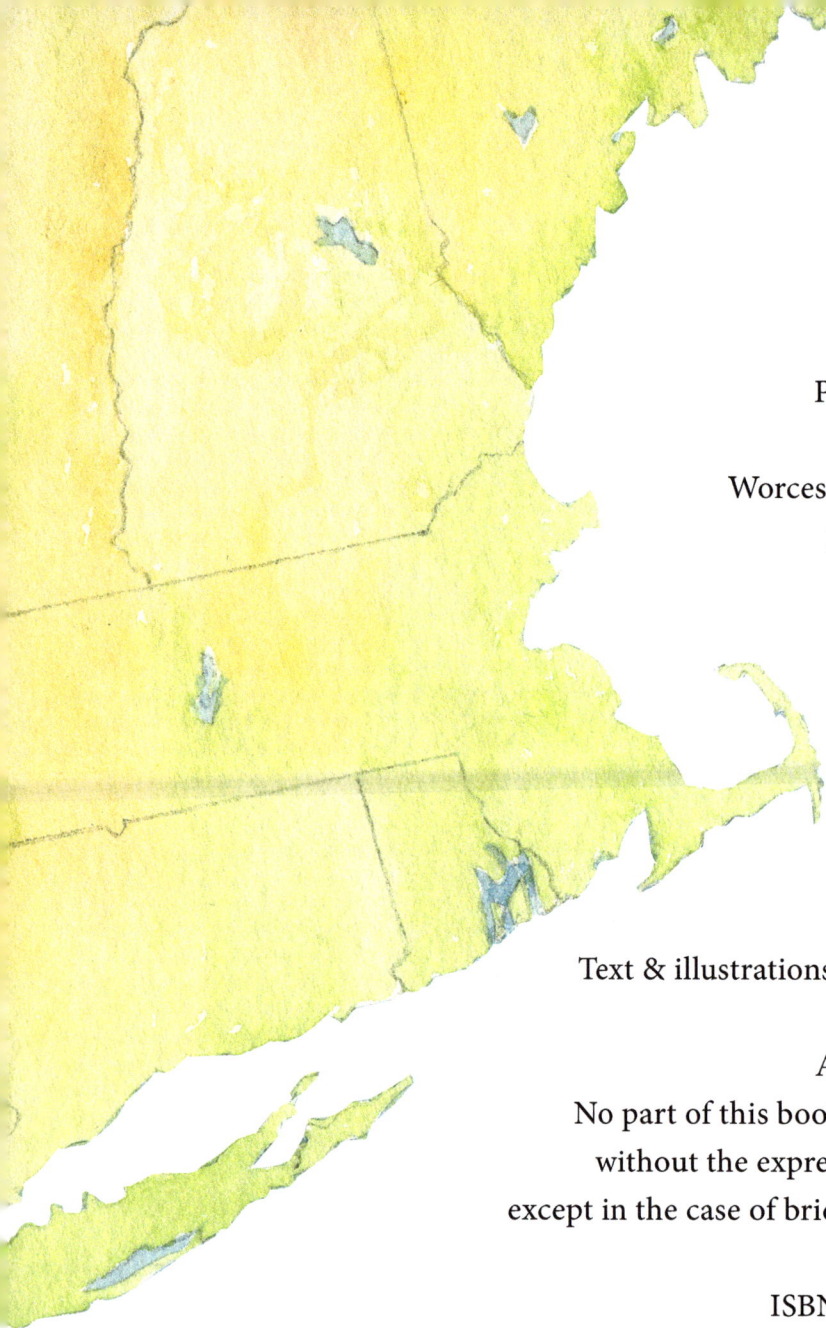

Published by
Pickerel Publishing
25 Arbutus Rd.
Worcester, Massachusetts 01606
jroachevans.com

ISBN: 978-1-7341532-0-0
Library of Congress Control Number: 2020902673

for the Gaskell boys

Nicholas, Jason,
& Christopher

with love

You may have heard
of a place
called the beach.

Perhaps you have been to the beach
and found seashells
or
seaweed there?

But are you aware of the many
creatures (marine animals)
that can be found on the beach?

You might be amazed to discover the fascinating marine animals that live at the seashore.

Sometimes you just find the shells of these creatures, but sometimes you can find them alive!

They may live on the rocks or under the sea.

There might be some in the sand beneath your feet!

Clams in the sand.

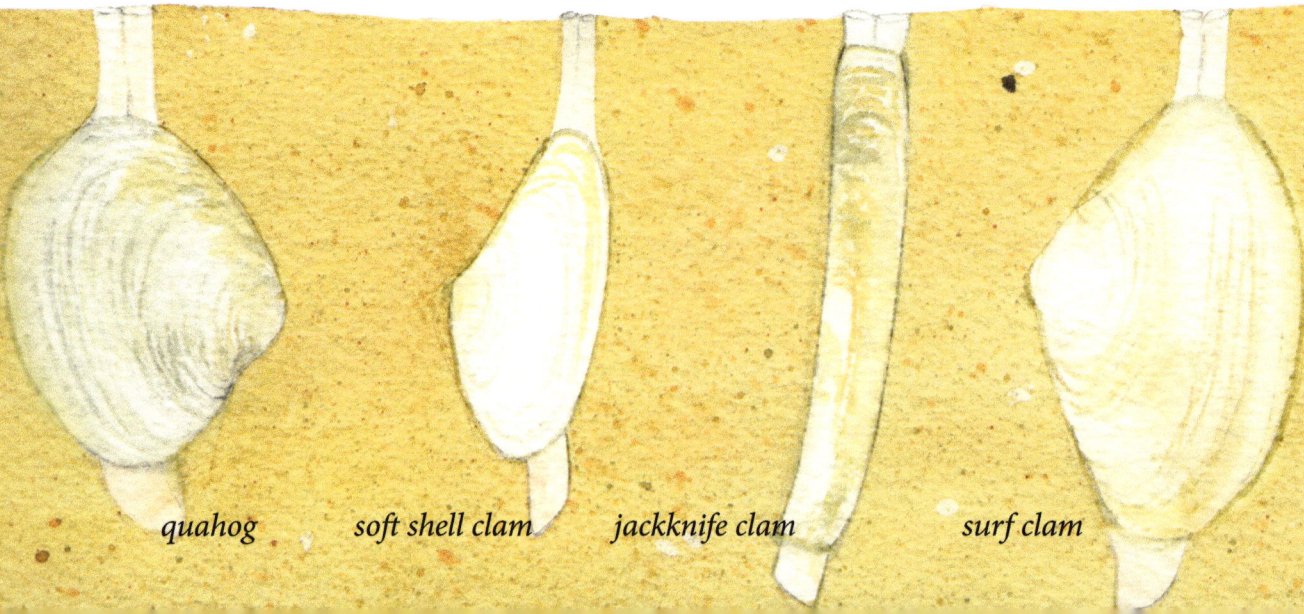

quahog   *soft shell clam*   *jackknife clam*   *surf clam*

Some creatures leave little trails in the sand.

Can you guess what little creature
makes these trails?

It's a **perwinkle snail** that makes
these little trails!

This snail's shell has a different pattern than the usual
periwinkle. The typical stripes of this periwinkle have
been worn and pitted to create this amazing design!

Periwinkle snails are marine mollusks.

They are often found on rocks
where they graze on blue-green algae.

You may also find a snail called a **dog whelk**.
Dog whelks are also called dogwinkles.
They are around the same size as the
periwinkles, but they like to eat
barnacles and blue mussels.

*The dog whelk's shell
is not as round as the
periwinkle's shell and
can be many different
colors.*

Here you can see periwinkles have left trails in the sand and sand trails on the rocks!

Many creatures live on the rocks with the periwinkles and the dog whelks. You might see a snail called a **limpet** that also feeds on algae. This one is a **tortoiseshell limpet** and it has a striped shell like a little hat.

There are also many **acorn barnacles**.

Acorn barnacles filter food from the water as it washes over them. When the tide is low they close their shells so they don't dry out! Barnacles are actually closely related to crabs because they are both crustaceans, even though they look very different!

Mussels also filter food from the water as they hang on to the rocks with sticky threads called beards! Can you find the one hiding in this painting?

Tortoiseshell Limpet

Barnacles

Rockweed/
Bladderwrack

Periwinkle

There are many types of crabs on the northeast coast.
One of the most unique crabs does not make its own shell.
These crabs are called **hermit crabs**.

Hermit crabs can be found in shallow water and tide pools.

Unlike other crabs, hermit crabs use the empty shells of periwinkles,
dog whelks, and moon snails for their homes. They can hide inside
when they get scared!
They have small back legs that help them to grip the shell and carry it
with them as they move around. It's like their mobile home!

As a hermit crab grows, they find bigger shells to move into. This hermit crab has a periwinkle shell for a home, but may need to move into a moon snail shell next.

This hermit crab is shown larger than actual size. The periwinkle shell is only about 1 inch long.

one inch

rock crab

These are what scientists call true crabs.
They have eight walking legs and two claws.
There are many different types of true crabs.
Three common true crabs are the **rock crab**,
the **green crab**, and the **spider crab**.

You may not see a live crab.
You might just find the shell
(called a carapace)
 on the beach.

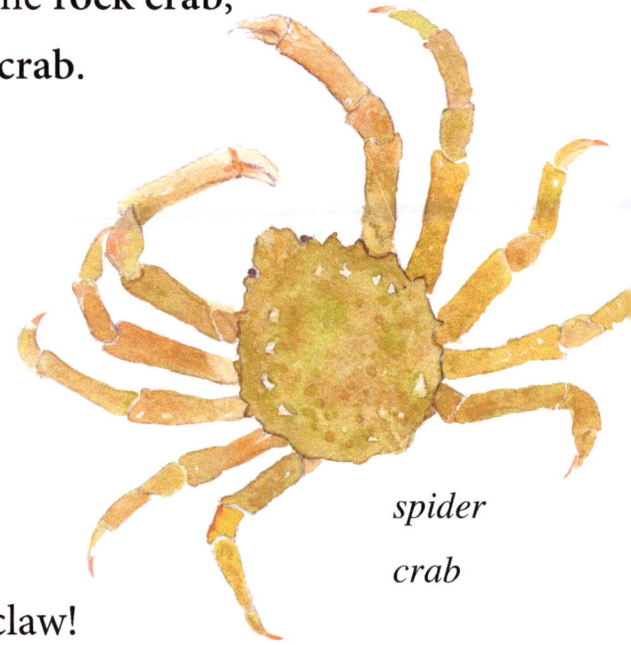

spider crab

Or you might just find a claw!

green crab

One of the largest
and most remarkable
of all the crabs is the
horseshoe crab.

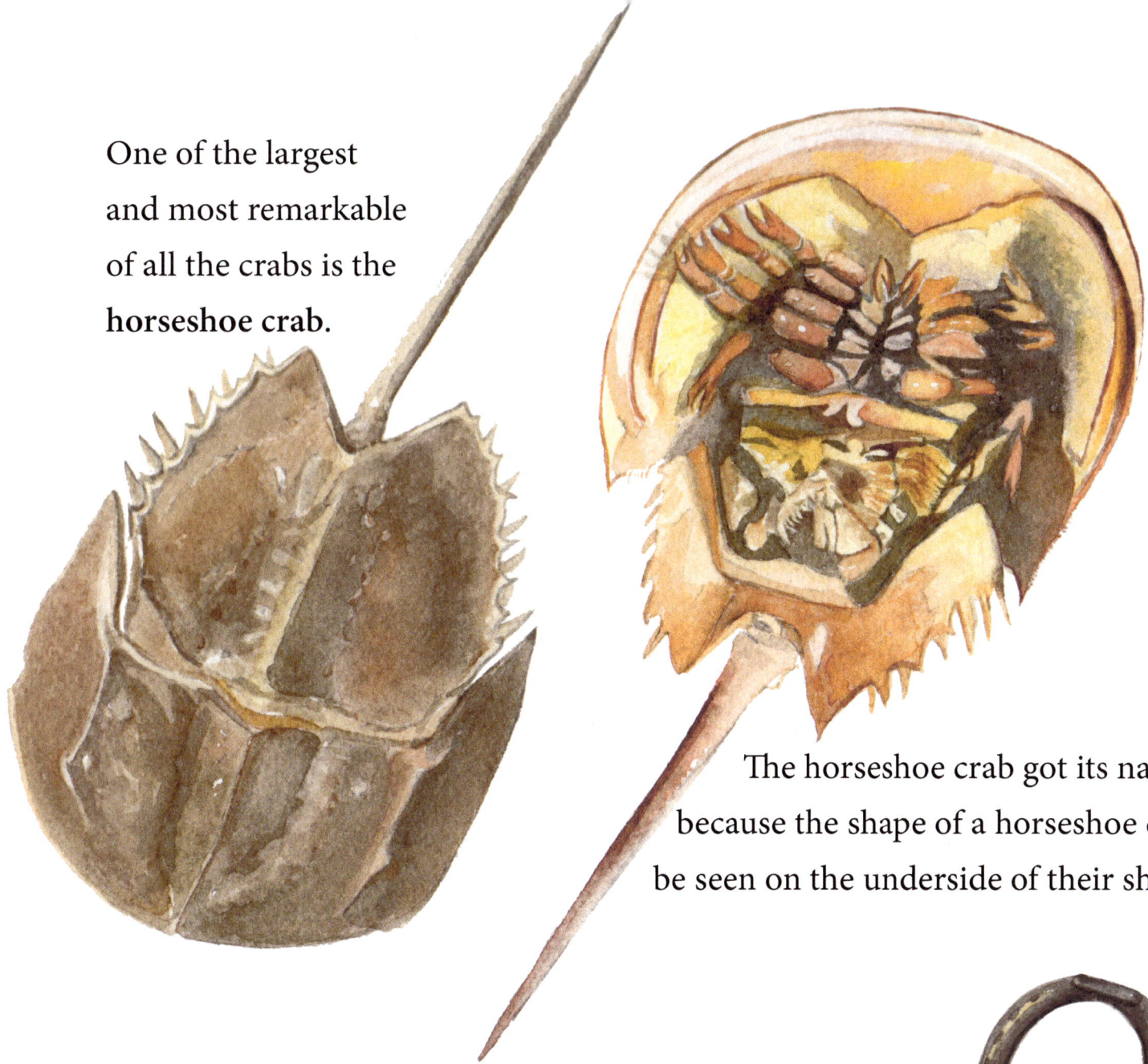

The horseshoe crab got its name
because the shape of a horseshoe can
be seen on the underside of their shell.

Horseshoe crabs have lived on this Earth for a very long time.
They are ancient creatures that lived alongside dinosaurs!

horseshoe

During the day, crabs usually try to hide from predators like seagulls. At night, crabs will come out of hiding to look for food (called foraging).

You may find a green or rock crab hiding in a tide pool, as long as a seagull hasn't spotted it first!

Seagulls love crab!

Another cool tide pool resident
is the **green sea urchin**.

You may not see a live sea urchin,
you may just find their beautiful,
round, pin cushion like shells
(called tests).

Their shells are truly works of art!

Alive with spines

and without spines

Sand dollars got their name

because they resemble silver one dollar coins.

Like its cousin the sea urchin,
the **sand dollar**
is also round and beautiful.

The shell of the sand dollar
has an artistic pattern;
it's shaped like a star
made out of flower petals.

The sand dollar
is a special treasure
for many shell collectors.

*A live sand dollar has a covering of soft
spines, hair (called cilia), and little tube feet
around and under its shell.*

One more cousin of the sand dollar and the sea urchin
is the **sea star.** The sea star is easy to identify because of
its star shape.

Like the sand dollar and the sea urchin,
it has little tube feet that help it move around.
What's unique about the sea star is that its feet
are on the underside of its arms!

You can find the sea star attached to rocks
alongside seaweed or in tide pools.

Like the sand dollar and the sea urchin,
sea stars cannot survive out of the water for long.

Please don't take them out of the water -
observe them where they are!

There are so many creatures to be discovered
on the beach and around the rocks.

It's fun to explore the tide pools
along the seashore.

You just might find
one of these fascinating
marine animals alive!

Just remember:
if they are alive
let them be.

In order to survive,
these creatures need to be
where they belong -

in or near the sea!

MOLLUSKS

Gastropods (Univalves) & Bivalves

Periwinkle

Dog Whelk

Tortoiseshell Limpet

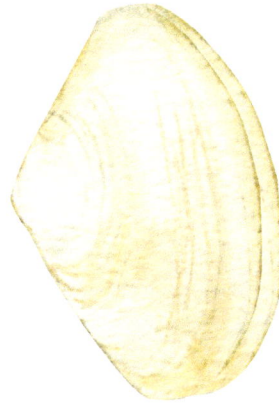

Surf Clam

Blue Mussel

ECHINODERMS

Sand Dollars, Sea Urchins, Sea Stars

Sand Dollar

Sea Urchin

Sea Star

CRUSTACEANS

Hermit Crabs, True Crabs, Horseshoe Crabs, & Barnacles

Hermit Crab

Acorn
Barnacle

Atlantic Rock Crab

Spider Crab

European
Green Crab

Horseshoe Crab

Atlantic Ocean

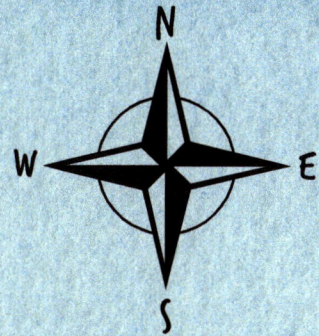

To learn more about marine animals,
check out these great books and you tube channel:

Seashells in My Pocket, by Judith Hansen,

The Seaside Naturalist by Deborah A. Coulombe, &

www.youtube.com   seacoastsciencectr

You can also check out Joanne's blog & YouTube videos
@ jroachevans.com

York Beach, Maine © W. Evans

Joanne Roach-Evans is the author and illustrator of several seashore books: *Seashells, Treasures from the Northeast Coast*; *Seaweed, Marine Algae from the Northeast Coast*; and *Marine Animals from the Northeast Coast*. Joanne grew up going to many of the beaches on the New England coast. When she was a kid, one of her favorite pastimes was fishing for crabs off the jetties at Sand Hill Cove in Rhode Island.

*If you've enjoyed this book please leave a review on Amazon!

www.ingramcontent.com/pod-product-compliance
Lightning Source LLC
Chambersburg PA
CBHW060835270326
41933CB00002B/93